AF882681

Visit Susie and her Animal Friends at
KidsKyngdom.com

Visit Susie and her Animal Friends at
KidsKyngdom.com

For my baby boy
Cameron

I hear You **Roar**!
Thank YOU for being YOU

Visit Susie and her Animal Friends at
KidsKyngdom.com

A fantastic thank you
to **Freddie**

She reviewed this book
and said she loved when
Susie and Bash hug the
Lion!

Visit Susie and her Animal Friends at
KidsKyngdom.com

Thank you for supporting Kids Kyngdom

Get your **FREE** audio book at KidsKyngdom.com

Visit Susie and her Animal Friends at
KidsKyngdom.com

Send Dr. Susie a Creative Message on Instagram

I may pick You!

Instagram

Visit Susie and her Animal Friends at
KidsKyngdom.com

Hi

I'm Susie

Visit Susie and her Animal Friends at
KidsKyngdom.com

Susie

to be specific

Visit Susie and her Animal Friends at
KidsKyngdom.com

You can call me

Dr. Sue

Visit Susie and her Animal Friends at
KidsKyngdom.com

This is my friend Bash

He's a Frog

Visit Susie and her Animal Friends at
KidsKyngdom.com

We crossed the **Atlantic**

Ocean and landed in **Africa**

Visit Susie and her Animal Friends at
KidsKyngdom.com

Shhhh
Do You Hear that?

You don't?

Visit Susie and her Animal Friends at
KidsKyngdom.com

Well, this Lioness does

She has amazing hearing

Visit Susie and her Animal Friends at
KidsKyngdom.com

She can hear things really far away

Like her **dinner**

Visit Susie and her Animal Friends at
KidsKyngdom.com

A lioness is a girl

Girl lions are way better hunters than boy lions

Visit Susie and her Animal Friends at
KidsKyngdom.com

When a girl lioness hunts

The boy lions stay at home to protect their Pride

Visit Susie and her Animal Friends at
KidsKyngdom.com

Pride is all the lions, lionesses and cubs in a family

Visit Susie and her Animal Friends at
KidsKyngdom.com

Cubs are the **babies**

Visit Susie and her Animal Friends at
KidsKyngdom.com

Protecting cubs is super important

Visit Susie and her Animal Friends at
KidsKyngdom.com

Lions will **Roar** to protect their Family

Visit Susie and her Animal Friends at
KidsKyngdom.com

Shhhhhhh

Listen closely

Visit Susie and her Animal Friends at
KidsKyngdom.com

They also **meow**

And they **hissssss**

Visit Susie and her Animal Friends at
KidsKyngdom.com

Just like cats

But way louder

Visit Susie and her Animal Friends at
KidsKyngdom.com

That's right

Lions are Cats

Visit Susie and her Animal Friends at
KidsKyngdom.com

Boy lions grow long hair

around their neck

Visit Susie and her Animal Friends at
KidsKyngdom.com

Their neck hair is called

a **mane**

Visit Susie and her Animal Friends at
KidsKyngdom.com

Not **rain**

Visit Susie and her Animal Friends at
KidsKyngdom.com

Visit Susie and her Animal Friends at
KidsKyngdom.com

Do you know the swahili word for Lion?

That's right! It's Simba

Visit Susie and her Animal Friends at
KidsKyngdom.com

Simba's can be so lazy

Sometimes they sleep all day

Visit Susie and her Animal Friends at
KidsKyngdom.com

Hey Bash

Can you sleep all day?

Visit Susie and her Animal Friends at
KidsKyngdom.com

Skip it?

Ribbit

Visit Susie and her Animal Friends at
KidsKyngdom.com

Come on Kyngs!

Visit Susie and her Animal Friends at
KidsKyngdom.com

Let's skip sleeping all day

Visit Susie and her Animal Friends at
KidsKyngdom.com

We have so many more animals to meet

Visit Susie and her Animal Friends at
KidsKyngdom.com

Come on Bash

Visit Susie and her Animal Friends at
KidsKyngdom.com

Let's go visit

a **Silverback** Gorilla

Visit Susie and her Animal Friends at
KidsKyngdom.com

See you later Kyngs

We'll meet you in the African Mountains

Visit Susie and her Animal Friends at
KidsKyngdom.com

Do you know how to tell Gorillas that you are friendly?

Let's learn about it with Dr. Susie and Bash!

Visit Susie and her Animal Friends at
KidsKyngdom.com

Sammie Kyng Loves Animals

Sammie is a lover of nature and animals

Her hobby is learning **all about animals**

Sammie is a proud Mom of 3 **Non-Furry humans** and is the creator of Dr. Susie Animal Safari

Susie is a young scientist. She is also ***Sammie's alter ego***

Meet the Gang at:

Kidskyngdom.Com

Instagram

Visit Susie and her Animal Friends at
KidsKyngdom.com

Copyright© MMXXI Sammie Kyng

All Rights Reserved

Request permission to reproduce ***more than a short quote*** of any part of this book at:

Kidskyngdom.com

Instagram.com/Kids_Kyngdom

ISBN: 978-1-959501-04-6

Published by: Kyngdom, LLC

Visit Susie and her Animal Friends at
KidsKyngdom.com

Did you love learning about **Lions** with Susie and Bash?

Please Share Your Review

Visit Susie and her Animal Friends at
KidsKyngdom.com

Made in the USA
Monee, IL
05 November 2026

67377490R00026